ANIMAL HOMES

Diane James & Sara Lynn

Illustrated by Sue Cony

TWO-CAN

Home and Dry

What is your home like? Animals have homes, too. They need shelter from the weather and a place to keep their families safe. Some even carry their homes about on their backs!

Many animals live on or under the ground. Others live high up in trees or rocks. Burrows, nests and hollows can all be animal homes.

GRIZZLY BEARS

Many grizzly bears live in very cold countries. In winter, they look for warm, dry dens.

Some grizzlies make their dens in caves. Others dig holes in the ground or shelter under trees.

Grizzlies catch fish to eat and climb trees to find honey. They eat plenty of plants and other food in summer because food is scarce in winter.

Bear cubs are born in the middle of winter. The mother grizzly and her babies sleep in their den until spring.

MOLES

Moles live in underground burrows where the soil is soft. They often make their home in people's gardens!

Moles use their strong front legs to dig. They have very sharp claws.

Moles cannot see well. They have whiskers which help them feel their way through tunnels and burrows.

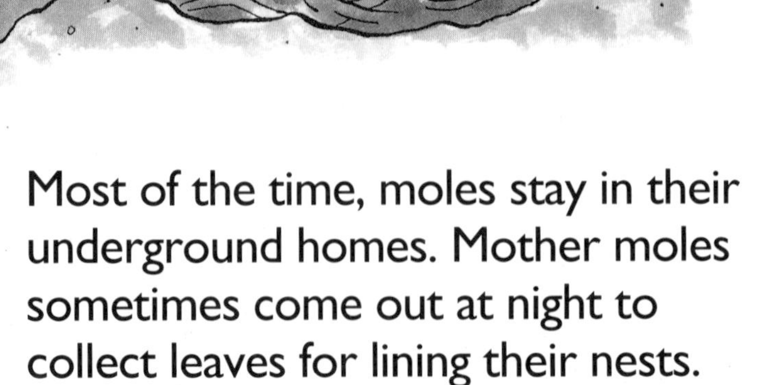

Most of the time, moles stay in their underground homes. Mother moles sometimes come out at night to collect leaves for lining their nests.

HARVEST MICE

Harvest mice build their nests in tall grasses. They use their long tails to help them climb.

Harvest mice weave strips of leaves round the stems of the grasses. This makes the frame for the nest. Then they line it with a bed of soft grass.

A harvest mouse's nest has to be strong enough to hold up to eight baby mice.

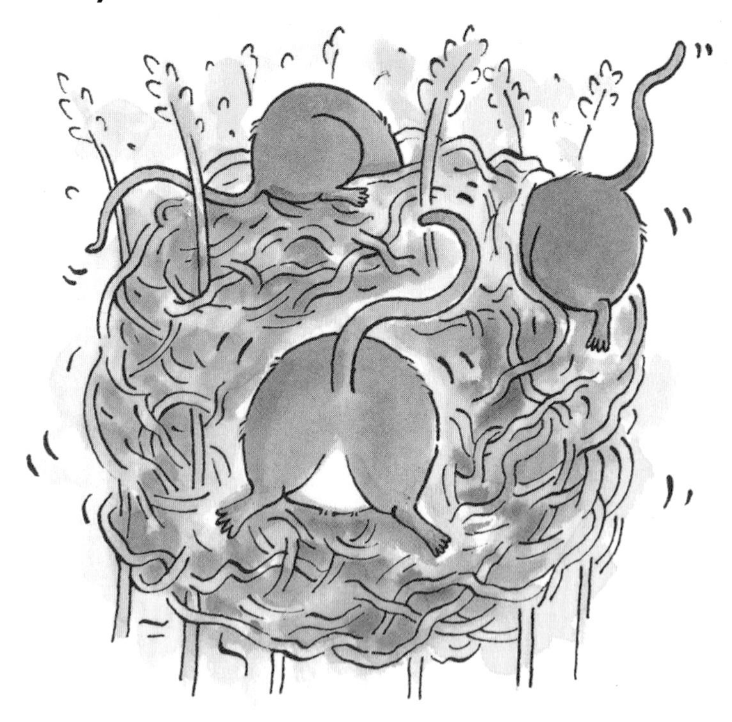

There is no special entrance to the nest. The mice just push in anywhere.

PRAIRIE DOGS

Prairie dogs are very friendly animals. They often live together in big groups. They make their homes in burrows under the ground.

If they sense danger, they bark to warn others. Then they jump quickly back into their holes to escape.

During the day, prairie dogs leave their burrows to eat grass and plants.

Prairie dogs say 'hello' to each other by pressing their noses together.

SNAILS

Snails take their homes with them wherever they go! They live in shells which they carry on their backs.

Snails like wet weather. They often come out to eat plants after it has rained. When it is dry, they stay in their shells and seal up the entrance.

Snails have soft, slimy bodies that leave a silvery trail as they move along. Their eyes are on long stalks.

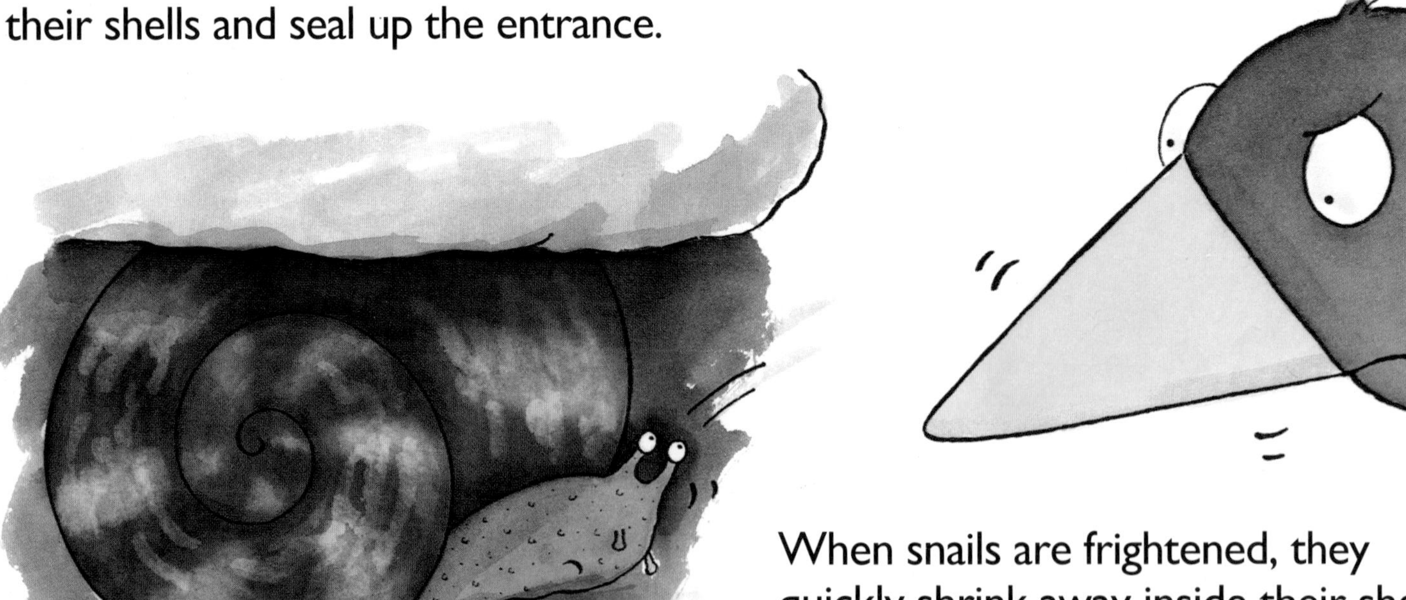

When snails are frightened, they quickly shrink away inside their shells. They hide under stones and wood where birds cannot find them.

BEAVERS

Beavers live in homes called lodges. They make their lodges from the branches of trees.

Beavers have very strong teeth. They use their teeth to chew bark and cut down trees.

Beavers build their lodges in the middle of a lake or river. The underwater entrance leads up to a large room.

Young beavers stay dry and warm in the lodge. Their parents bring them plants, twigs and leaves to eat.

RABBITS

Rabbits live in big groups. They often share their underground tunnels and burrows. They join them together to make a big home called a warren.

Rabbits have long ears to pick up faint sounds. They also have a good sense of smell. Their noses twitch as they sniff out danger.

Rabbits use their front paws to dig holes and tunnels. Their back legs are longer and stronger, for hopping.

When they are afraid, rabbits hop back into their burrows for shelter.

TREE SQUIRRELS

There are lots of different kinds of squirrels. The ones that live in trees make their homes in hollow trunks or among the branches.

Tree squirrels often have two homes. One is a warm nest lined with leaves and bark. The other is a cooler, more airy nest for hot days.

Squirrels spend much of their time looking for nuts and berries to eat.

They hide the food away in their nests or bury it underground so that no-one else can find their store!

WASPS

Wasps are insects which live in nests. They often work in big groups and help each other to build their nests.

Some wasps build their nests from thin layers of paper. They make the paper by chewing up plants and old wood. Other wasps make mud nests.

Female wasps sleep through winter. In spring, they start a new nest where their babies will live and grow.

Wasps usually eat the nectar from flowers. Sometimes they catch other insects for their babies to eat.

QUIZ

What is special about a snail's home?

What is a harvest mouse's nest made from?

What do prairie dogs do if they sense danger?

Where do grizzly bears sleep in the winter?

What is a beaver's home called?

Where do rabbits go when they are frightened?

What do moles use to dig their holes?

How many nests does a tree squirrel have?

INDEX

If you have enjoyed this book, look out for the full JUMP! STARTS range:

PLAY & DISCOVER • What We Eat • Rain & Shine
CRAFT • Play with Paint • Play with Paper • Fun Food • Dress Up
ANIMALS • Pets • On Safari • Underwater • On the Farm • Animal Homes • Birds

For more information about TWO-CAN books write to:
TWO-CAN Publishing Ltd, 346 Old Street, London EC1V 9NQ

First published in Great Britain in 1993 by
Two-Can Publishing Ltd, 346 Old Street, London EC1V 9NQ
in association with Scholastic Publications Ltd.

Printed and bound in Hong Kong 2 4 6 8 10 9 7 5 3 1

A catalogue record for this book is available from the British Library.

Pbk ISBN 1-85434-205-3
Hbk ISBN 1-85434-200-2

Photo credits: pp2-3 Bruce Coleman, p5 Bruce Coleman, p7 Bruce Coleman, p9 Bruce Coleman, p11 Survival Anglia, p13 Oxford Scientific Films, p15 Bruce Coleman, p17 Bruce Coleman, p19 Survival Anglia, p21 Survival Anglia

Editor: Lucy Duke Designer: Beth Aves